PROJECT SCIENCE

FORCES AND ENERGY

Alan Ward

Franklin Watts

New York • London • Toronto • Sydney

© 1992 Franklin Watts

First published in the United States
in 1992 by Franklin Watts, Inc.

Library of Congress Cataloging-in-Publications Data

Ward, Alan.
    Forces and energy / by Alan Ward.
        p.   cm. (Project science).
    Includes index.
    Summary: Demonstrates the principles of different types of forces
and energy through simple experiments.
    ISBN 0-531-14132-2
    1. Force and energy — Experiments — Juvenile literature. [1. Force
and energy — Experiments. 2. Experiments.] I. Title.
QC73.4.W37   1992
531.6——dc20   91-3731 CIP AC

Series Editor : A. Patricia Sechi
Design : Mike Snell
Illustrations : Michael Lye
Typesetting : Spectrum, London

Printed in Great Britain

CONTENTS

WHAT ARE FORCES?

Forces can make an object start moving or they can make it move faster. They can make an object change direction, slow down or just stand still. Forces can even make an object change its shape. We cannot see forces, but we can see their effects on objects around us.

Pushes and pulls

Have you ever watched a caterpillar or an earthworm moving along? Both these creatures move by pushing out the front half of their body, and then pulling in the back half. Pushes and pulls are two kinds of forces.

Stand a short distance away from a friend and face toward each other. Lean forward slightly and place your hands on your friend's shoulders. Ask your friend to do the same as you. As long as you both push each other's shoulders, neither of you will fall over.

Now stand closer together so that your toes are touching your friend's toes. Hold each other's wrists firmly and lean backward gently. As long as you both pull each other, neither of you will fall over.

Whenever there is a force in one direction, an equal force is needed in the opposite direction to stop things from falling over or collapsing.

Twists

Apart from pushes and pulls, what are the other kinds of forces?

YOU NEED:
- a piece of cardboard 8 cm × 20 cm (3 in × 8 in)
- two rubber bands
- a sharp pencil

Use the pencil to punch a hole at each end of the cardboard. Now thread a rubber band through each hole and loop one end back through the rubber band.

Slip one rubber band over a door handle or hook. Pull the other rubber band with one hand while you twist the cardboard around with the other hand.

Squeezes

Squeezes are a fourth type of force. You have to push in order to squeeze something into a smaller space. When you stop pushing, the squeezed object will pull and push itself back into shape.

What are the four different types of forces that we have just talked about? Can you name the force which is acting in these pictures?

When the rubber band is tight, let go of the cardboard and watch it spin. As the rubber band untwists, it forces the cardboard to spin around. Twists are a third type of force.

GRAVITY AND WEIGHT

Every substance is made up of tiny pieces of matter called particles. The smallest particles are called atoms, and they can join together to make molecules. Atoms and molecules are so small that we cannot see them separately.

Solids like wood, liquids like water, and gases like air are all made up of these particles. Anything that is made up of atoms and molecules is called matter.

The amount of matter in an object is called its mass. Mass is measured in kilograms or pounds. The mass of one object pulls on the mass of another object — this is the force called gravitation. The more mass there is, the greater the pull. The pull, or gravitation, of the earth is simply called gravity.

When an apple falls off a tree, both the apple and the earth are pulling each other. But you don't notice the apple's tiny pull because the earth is huge and so its pull is much bigger.

A trick with gravity

Place the large orange and the small orange on opposite ends of the tray. Now balance the tray on your hand. Take one of the oranges off the tray. What happens?

YOU NEED:

- a large orange
- a small orange
- a tray

Watching your weight

The force of gravity that pulls on the mass of your body is called weight. When you try to resist the force of gravity you feel "weight." Because we live on the earth, our bodies have to resist the force of gravity all the time and so they feel weight.

Did you know?
Astronauts feel weightless in space, but the mass of their bodies is the same as it would be on earth. Only when the engines of their spacecraft start pushing them along do they have a sense of weight.

Every time you eat or drink you are adding more mass to your body. This increases the pull of gravity, so your body has to use more force to resist gravity. You feel the extra force as extra weight.

You can find out whether you have gained or lost mass by standing on a bathroom scale. This will show your weight. Try weighing yourself before and then after eating a large meal. Did your weight change? If so, did your body gain or lose mass?

THE FORCE OF GRAVITY

YOU NEED:

- a shoebox or similar cardboard box
- a heavy object, such as a stone or small brick or some modeling clay

Gravitation acts wherever there is mass.

Place the box in the middle of a table. The box will not easily fall off the edge of the table. It is stable.

Now move the box to the edge of the table until about half of it is sticking over the table edge. You can easily knock the box off the table now. The box is unstable.

What do you think will happen if you place a heavy object inside the part of the box sitting on the table? Will it be so easy to knock the box off now?

Resisting gravity

Blow up the balloon and tie a knot in the end. Hold

YOU NEED:

- a small balloon
- a bendable straw

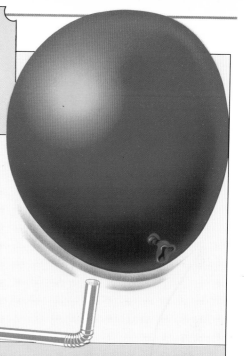

the balloon above the short end of the bendable straw and blow through the straw. If you let go of the balloon, it will hover in the air.

What has happened?

The push of the air from the straw resists the pull of gravity which acts on the balloon. When you stop blowing, the balloon will fall to the ground. The earth's gravity will pull it down.

YOU NEED:

- a round flat container, such as an empty can
- modeling clay
- a piece of cardboard, about 25 cm × 20 cm (10 in × 8 in)
- a pile of books

Rolling uphill

Did you think that objects can only roll downhill? Find out if you can make a circular object roll up a slope.

Stick a lump of modeling clay inside the can. Place the pile of books on a flat surface and make a ramp by leaning the piece of cardboard against the books. Now try and roll the can up the ramp. You may have to change the position of the clay a few times to make the can move.

Do you know why the can rolls uphill? Does it still move if you take away the modeling clay?

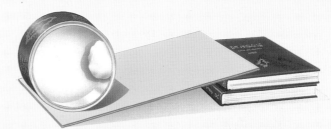

What has happened?

The pull of gravitation is concentrated where there is heavy mass (the clay) and where the box is being pushed or supported by the table. This pull stops the box from tipping over the edge of the table. In the same way, gravitation pulls a heavy mass (the lump of modeling clay) up a ramp.

FRICTION

Friction is the force that acts against, or resists, movement. It happens when two surfaces rub against each other. Two rough surfaces will cause more friction than two smooth surfaces.

Testing for friction

Try these tests to find out about the different amount of friction caused by rough surfaces and by smooth surfaces.

Make a slope by resting one end of the large mat against the pile of books. The smooth side of the mat should face upward.

Now try sliding the small mat down the slope, first with the smooth side facing up and then with the rough side facing up. Does it slide down easily both times?

Turn the large mat upside down so that the rough side is facing upward.

YOU NEED:

- a small table mat
- a large table mat (The mats should be smooth on one side and rough on the other side.)
- a pile of books

Now slide the small mat down the slope as you did before.

Do you know what happens when there is too little friction between surfaces?

What has happened?
When the two smooth surfaces are in contact with each other, there is less friction and so less force is needed to make the mat slide. When the two rough surfaces are in contact, more friction is present and so more force is needed to make the mat slide. Rough surfaces have more resistance to movement, and so they cause more friction.

Brakes use friction

The brakes on a bicycle work by friction. When you apply the brakes, small pads of rubber grip the metal rim of your bicycle wheel. The force of the friction between the pads and the rim makes the bike slow down.

Friction also causes heat. Next time you brake hard, touch the brake pads immediately afterward. They will be hot. The friction between the brakes and the wheels has created heat.

When you rub your hands together, they are causing friction. Can you feel heat?

Did you know?

A meteor is a rock from outer space. It travels through space at a speed of thousands of miles per hour. When the meteor enters the earth's atmosphere it encounters friction with the air. The meteor glows and starts to burn.

PRESSURE

If you try to walk on fresh snow, you start sinking down into the snow. All the force of your weight is acting through your shoes and pressing down on the snow.

If your shoes were wider and longer, they would press down on a bigger area of snow. Each part of the snow's surface under your shoes would receive only a small share of your weight, and so you would not sink into the snow. People who have to walk on deep, fresh snow often wear special snowshoes.

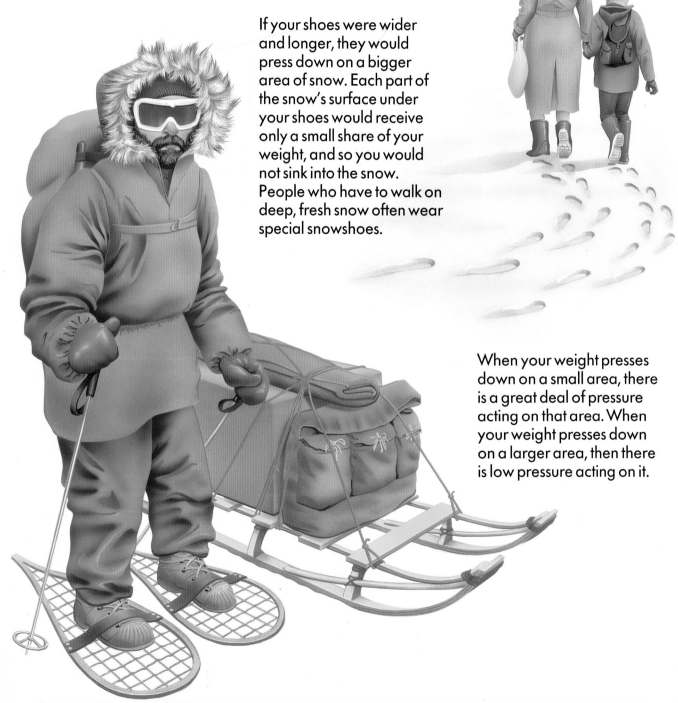

When your weight presses down on a small area, there is a great deal of pressure acting on that area. When your weight presses down on a larger area, then there is low pressure acting on it.

Snowshoes reduce pressure. Do you think a thumbtack reduces or increases pressure? It has a large surface on which you can press down, but a tiny surface to make a hole. If you exert a small force on a thumbtack, there is enough pressure to make a hole in a piece of wood.

A magician on a bed of nails

You may have seen pictures of people lying on a bed of nails. Do you know how they can do this?

The force of the person's weight is spread over and shared by lots of nails. Each nail point only has to support a small part of this force. The magician must climb very carefully onto the bed of nails. He makes sure that his whole body is always supported by enough nails to stop any one nail from hurting him.

Did you know?

Air takes up space and so it has mass. The air around the earth is pulled by the force of the earth's gravity. The weight of the air pressing down on you could be as much as the weight of two elephants. The force is shared by all of your body, so you are not hurt. Also, the air inside your body presses outward to balance the pressure of air outside.

WHAT IS ENERGY?

When you throw a flat stone to make it skip across the surface of water, you are using energy. You give the stone a push and it flies out of your hand. When a force makes an object move, we say that the force does work. Energy is what makes things work.

Energy is mysterious. You cannot see it, but you can see what it does to things around you.

Storing energy

Energy can be stored and used later. Stored energy is called potential energy. It is one of the two main forms of energy.

Cut two small slots, 2 cm (1 in) apart, at each end of the cardboard.

YOU NEED:

- a piece of stiff cardboard, about 7 cm × 10 cm (3 in × 4 in)
- scissors
- a rubber band
- colored pens

Did you know?

Your body burns the substances contained in food. These substances act as a sort of fuel to keep you alive and active. If the energy-supplying food that you eat is not used by your body, it may turn into unwanted fat. So if you just sit around reading this book, without doing the activities, the energy you get from your food will be wasted!

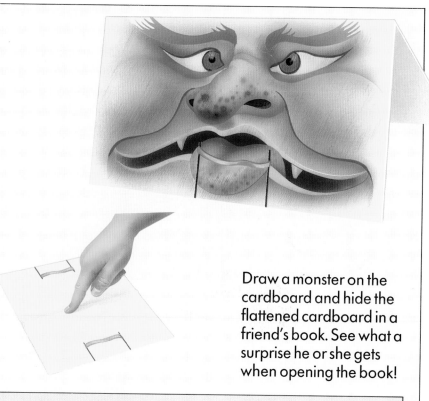

Fold the cardboard in half. Loop the rubber band through the four slots.

Press down on the cardboard so that it is flat and the rubber band is stretched. What happens when you take your hand off the cardboard?

Draw a monster on the cardboard and hide the flattened cardboard in a friend's book. See what a surprise he or she gets when opening the book!

What has happened?

When you press the cardboard flat, you use energy to stretch out the rubber band. Some of the energy is stored in the rubber band. When you take your hand off the cardboard, that energy is set free and it makes the cardboard jump up.

The energy stored in the rubber band is potential energy. It is energy waiting to be used. The energy that humans store up from food is also potential energy.

Do you remember the twisting cardboard that you made earlier (p. 6-7)? How did it work? Energy made the cardboard spin around. Where was that energy stored?

KINETIC ENERGY

Energy is never made or lost — it is changed from one kind of energy to another. When we store energy as potential energy and then set it free, the energy that is set free is called kinetic energy. It is the energy of movement.

Superbubble blower

Carefully cut the dishwashing liquid bottle in half. Take the top half of the bottle and cut slits 1 cm (½ in) deep, around the edge of the cut end. The slits should be about 2 cm (1 in) apart.

Bend the plastic between the slits to make a flower-like structure. This is your superbubble blower.

Stir a small amount of dishwashing liquid into a shallow bowl of warm water. Dip the end with the slits into the water. Lift the blower out of the water and point it downward while you blow through the top.

You should be able to make a bubble as large as a dinner plate. Keep your hand over the top of the blower to stop air from

escaping. Now take your hand away from the top of the blower. Can you feel the gentle jet of air being forced out as the bubble becomes smaller?

Take the sheet of paper and fold it diagonally to make a triangle. Fold it in half again to make a smaller triangle.

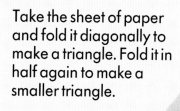

Open the paper and press down along the fold lines.

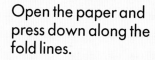

Stick a lump of modeling clay onto a flat surface and push the blunt end of the pencil into the clay.

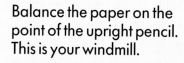

Balance the paper on the point of the upright pencil. This is your windmill.

Can you make the jet of air from the shrinking bubble spin your windmill?

What has happened?

Energy from the effort of blowing has made the bubble stretch. While it is stretched, the bubble stores some of that energy as potential energy. When you remove your hand from the top of the blower, the energy stored in the bubble forces the bubble to become smaller. The air is squeezed out of the bubble as a small jet of wind.

The potential energy in the filmy bubble is changed into the energy of movement, or kinetic energy. This kinetic energy makes the air move and spin the windmill.

Did you know?

A bubble is made of air that is squeezed inside a sort of "envelope." The bubble has an outside layer and a layer inside. These layers are made of detergent and water, and they are elastic, or stretchy. In between the two layers of the bubble is mostly water, which holds the bubble together.

ENERGY WAVES

Energy can travel in wavy or shaky lines. Heat and light energy from the sun travel through space as invisible, wavy waves. Some objects tremble or vibrate, making shaky waves in the air. When the waves reach our ears, we hear them as sounds.

Making waves

Tie the end of the rope to a leg of the chair or table. Flick the other end of the rope and watch what happens. Now move the rope up and down. What happens this time? You are making waves.

YOU NEED:

- a short rope
- a chair or a table

Fill the bowl with water. Float the cork in the middle and wait until the water is quite still. Touch the surface of the water with one finger. You will see small waves, or ripples, spreading out and bouncing off the sides of the bowl. What happens to the cork?

YOU NEED:

- a wide bowl
- water
- a cork

What has happened? When you force the rope to move, you give it energy. Each part of the rope only goes up and down, but something seems to be moving along the rope. This is a wave of energy.

The ripples on the water carry the energy that you put into the water when you touched it.

Another kind of wave motion

Place five marbles in a line on a flat surface. The marbles should be touching each other. Hold the first four marbles steady with your fingers. Flick another marble to hit the first marble in the line and watch what happens.

The flicked marble hits and pushes the first marble being held by your fingers. This creates a wave of energy. Each marble shakes and passes on the motion. But the last marble is free. When the energy reaches it, the last marble rolls away.

Can you make two marbles roll away?

Did you know?

Energy from the sun warms the air around us and makes it move. Breezes, winds, and storms are moving air. Winds touch and push the seas and oceans to make waves. The waves are like huge ripples that carry their energy to beaches far away.

HEAT ENERGY AND LIGHT ENERGY

The sun sends out energy in the form of heat and light. This energy travels as waves of light energy and heat energy. Energy from the sun is called solar energy.

YOU NEED:

- two cups
- cold water

Heat from the sun

Fill the cups with cold water. Put the same amount of water in each cup. Place one cup in bright sunshine, either on a windowsill or outside, and place the other cup in deep shade.

After half an hour, dip your fingers into each cup. Which cup of water feels warmer? This experiment will probably work better in summer.

The magic touch

Put the coins on the plate. Make sure they are cold. Close your eyes and ask a friend to choose a coin and to remember its date. Now ask the friend to hold the coin against the side of his or her forehead for 30 seconds.

Tell the friend to place the coin back on the plate. Open your eyes. Quickly touch all the coins. You will know which coin your friend chose because it will be warmer than the others. The coin was warmed by energy from your friend's body.

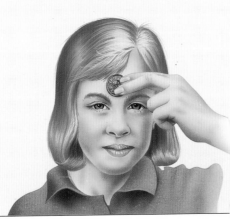

YOU NEED:

- five coins, each with a different date
- a plastic plate

Ask an adult if you can have two potted plants. Place them both outside. Put one in a warm position and where it will get plenty of light from the sun. Put the other plant in a very shady spot. Make sure you water both plants. What do you think will happen to them? Will both plants grow as well as each other? You will get a hint if you read this page again!

Food from the sun

Plants capture some of the light energy from the sun. A green chemical in their leaves uses the sun's energy to make food. The food is made from chemicals in water, soil, and air. Plants use some of this food for their own life and growth and store some of it.

Animals, including people, cannot use the sun's energy to make their food. Even factories cannot make food from sunlight. Only green plants can make food from sunlight. We have to eat food that comes from plants and from animals that have eaten plants, to keep us alive and make us grow.

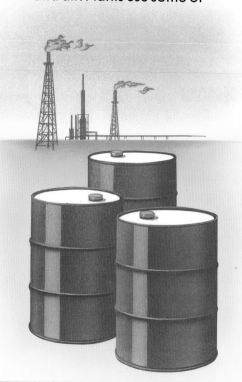

Did you know?

Millions of years ago, before there were people in the world, the living things on Earth were mostly plants and some animals. When these living things died, they rotted and formed coal and oil underground.

Coal and oil are fuels that we burn to provide energy for factories and homes. This is energy that once came from the sun. Many people are now wondering what we will do when all the coal and oil in the world are used up.

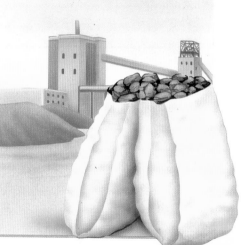

ENERGY FROM FLOWING WATER

Solar energy is energy from the sun. It heats water in puddles and in rivers, lakes, and seas. This heated water evaporates, or forms a gas, and rises high in the air.

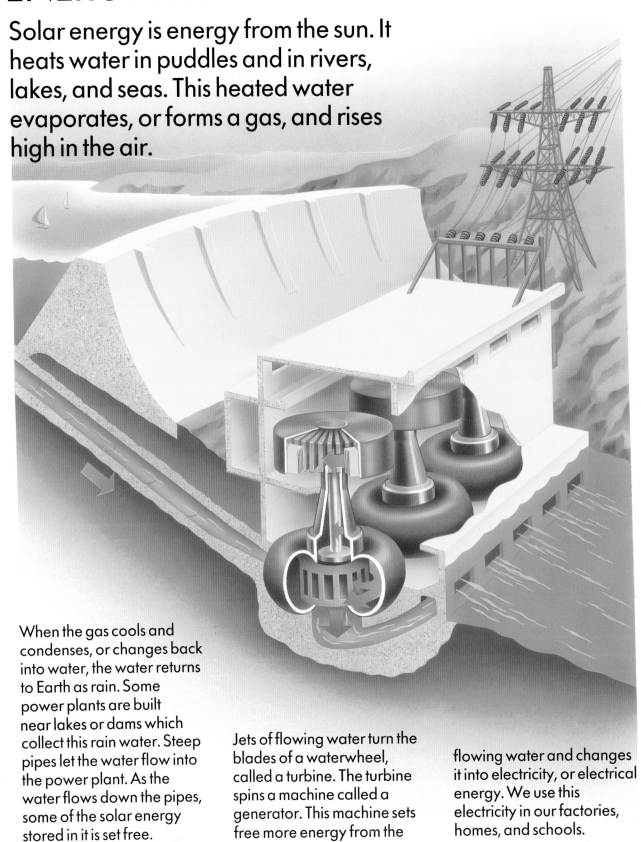

When the gas cools and condenses, or changes back into water, the water returns to Earth as rain. Some power plants are built near lakes or dams which collect this rain water. Steep pipes let the water flow into the power plant. As the water flows down the pipes, some of the solar energy stored in it is set free.

Jets of flowing water turn the blades of a waterwheel, called a turbine. The turbine spins a machine called a generator. This machine sets free more energy from the flowing water and changes it into electricity, or electrical energy. We use this electricity in our factories, homes, and schools.

Make a waterwheel of your own and test it in the kitchen sink.

Cut out four pieces of plastic, about 3cm × 5cm (1in × 2in). Mold the clay into a round lump around the center of the knitting needle. Push the pieces of plastic into the clay. Now hold your waterwheel under a running faucet and watch it turn.

YOU NEED:

- modeling clay
- a piece of stiff plastic
- scissors
- a knitting needle

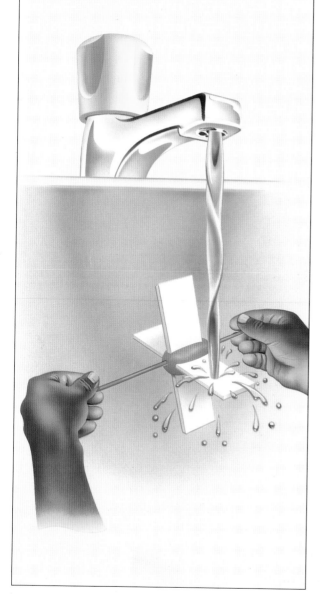

Did you know?
Solar energy is free. It may last forever, like water power. Power plants that set energy free by burning fuels, such as coal and oil, will be useless when all these fuels have been used up.

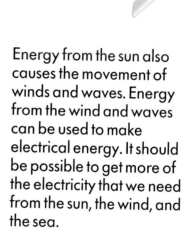

Energy from the sun also causes the movement of winds and waves. Energy from the wind and waves can be used to make electrical energy. It should be possible to get more of the electricity that we need from the sun, the wind, and the sea.

CHEMICAL ENERGY

When chemicals act on each other, the chemical energy inside them is set free. The booms and big bangs, whizzing rockets and dazzling colors of fireworks are all produced when energy stored in chemical powders inside the fireworks is released by burning.

When you pull a snapper, tiny amounts of chemicals react inside and release a safe explosion of energy.

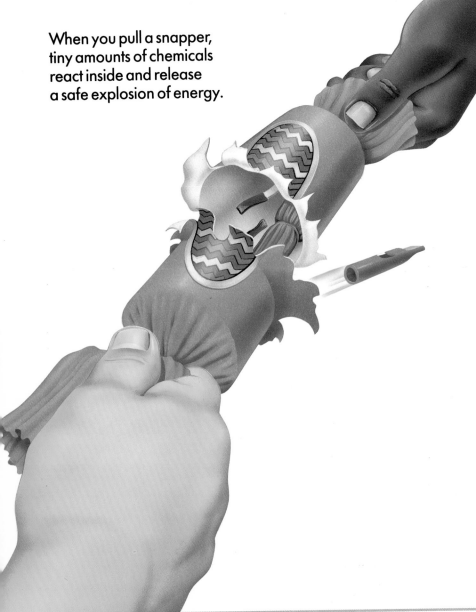

A bottle cannon

You can use chemical energy to shoot a safe missile. But take care to follow the instructions and do not point your missile at anyone!

You must be able to press in the stopper of the bottle, but make sure it does not fit too tightly. Put the bicarbonate of soda into the bottle. Pour in a little vinegar, then quickly push the stopper in the bottle.

Wait, and you will hear the gas fizzing inside. Then "POP!" the stopper shoots out of the bottle.

Dancing grapes

Fill the glass with soda and put the grapes into the glass. They will sink to the bottom and then rise again at once. They are lifted up by the bubbles of gas in the soda.

After about 10 minutes, some of the bubbles will disappear and the grapes will sink. But be patient, because more bubbles are forming, ready to make the grapes rise again. The grapes will keep sinking and rising in a kind of dance.

What has happened?

When bottled club soda is made in a factory, energy is used to force the gas to dissolve in the soda. The gas dissolves at high pressure. When you open a bottle, the pressure inside it becomes lower. The energy that was used to keep the gas dissolved is released. It is released in gas bubbles which make the grapes rise.

What has happened?

When the bicarbonate of soda reacts with the vinegar, chemical energy and gas is released. The pressure of the gas increases inside the bottle until it finally pushes the stopper out. As the stopper pops out, the stored energy is used up.

Did you know?

Bicarbonate of soda is one of the chemicals in baking powder. When baking powder is mixed with water and heated, energy is released from the powder to form gas bubbles. Baking powder is used in cake making to make cakes light and fluffy. When the cake mixture is baked in the oven, the gas bubbles from the baking powder make the mixture rise.

EXTRA PROJECTS

YOU NEED:
- a piece of paper about 12cm × 12cm (5 in × 5 in)
- scissors
- a pencil

Pushes and pulls

Fold the square of paper diagonally in half to make a triangle.

Fold it in half a second time to make a smaller triangle.

Make a cut in the paper as shown here. Pull forward the two "ears" you have just made and draw a face as shown.

Bend down a triangular flap on each side of the paper. You have now made a long-eared bat.

Hold the two flaps in one hand while you push and pull the paper tail with the other hand. Watch your long-eared bat as it flaps its ears.

Potential energy

Put a rubber band over the first two fingers of one hand. Pull the rubber band to make a loop. Now bend all four fingers and tuck them inside this loop in the same way that you make a fist. Open up your hand and the rubber band will jump onto your other two fingers.

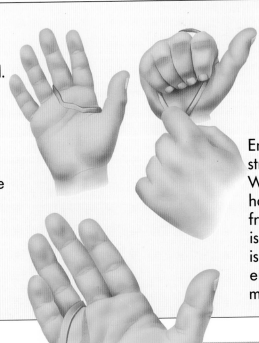

Energy is stored in the stretched rubber band. When you open up your hand you set this energy free. Stored energy, which is called potential energy, is changed into kinetic energy, or the energy of movement.

Heat energy

Rubber warms up when it is stretched. Pull a thick rubber band tightly between both hands. While it is stretched, place your lips against the rubber. How does it feel?

Now relax your hands so the rubber band is no longer stretched. How does it feel when you place your lips against it now? When one type of energy is changed into another kind of energy, heat is often given off. We cannot see this heat energy, except in things which glow red when hot, but we can feel it.

Hairpin walks

Pull the two sides of the hairpin slightly apart. Hold the ruler, edge upward, so that it is level with the table. Straddle the hairpin over the ruler so that the ends only just touch the table. Try to keep your hand still.

Mysteriously, the hairpin starts walking along the ruler! This happens because you cannot stop the muscles in your hand from twitching, or the blood being pumped around your fingers. The

YOU NEED:

- a hairpin
- a ruler

energy from these tiny movements causes the hairpin to move.

GLOSSARY

A

atom
The smallest part of an element; all atoms of an element are the same.

C

chemical energy
Energy which is set free when two or more chemicals act on each other or when chemicals are heated.

condense
To change into a liquid, as when a gas cools and turns into a liquid.

E

electricity
Energy that can flow along wires.

energy
The mysterious "go" in things which can be stored and then set free to do work.

evaporate
To turn into a gas, as when a liquid is heated and turns into a gas.

F

force
Any push, pull, twist or squeeze.

friction
The force that occurs when two surfaces rub against each other.

fuel
Any substance, such as coal or oil or wood, that can supply energy.

G

generator
A machine for changing energy into electrical energy or electricity.

gravitation
The pull that makes two different objects move toward each other.

gravity
The pull, or gravitation, of the earth.

H

heat
Energy that is supplied by atoms and molecules when they move backward and forward quickly.

K

kinetic energy
The energy of movement.

M

mass
The amount of matter found in any substance.

matter
Anything that takes up space and has mass; it is made up of atoms and molecules.

meteor
A rock which glows very brightly as it falls from outer space down to earth.

molecule
A very small piece of substance which is made when two or more atoms join together.

P

particle
A tiny piece of matter.

potential energy
Energy that is stored up and is waiting to be used.

power plant
A building or group of buildings in which electricity is produced, or generated.

pressure
The force which presses down on the surface of things.

S

solar energy
Energy from the sun.

stable
Not likely to fall; a stable object will not tip or fall over easily.

T

turbine
A wheel with blades that can be pushed and spun by a jet of water or steam. A turbine drives a generator in a power plant.

U

unstable
Likely to tip or fall; an unstable object will tip or fall over easily.

W

weight
The force of gravity which pulls on an object. We only feel weight when we resist this pull of gravity.

INDEX